Daniel S Troy

The Cosmic Law of Thermal Repulsion

Daniel S Troy

The Cosmic Law of Thermal Repulsion

ISBN/EAN: 9783744666381

Printed in Europe, USA, Canada, Australia, Japan

Cover: Foto ©berggeist007 / pixelio.de

More available books at **www.hansebooks.com**

THE COSMIC LAW

OF

THERMAL REPULSION

AN ESSAY SUGGESTED BY THE PROJECTION
OF A COMET'S TAIL.

Thermal Repulsion, like Gravitational Attraction, is universal between Masses as well as between Molecules of Matter.

NEW YORK:

JOHN WILEY & SONS,

15 Astor Place.

1889.

DRUMMOND & NEU,
Electrotypers,
1 to 7 Hague Street,
New York.

FERRIS BROS.,
Printers,
326 Pearl Street,
New York.

PREFACE.

THIS essay embodies ideas the development of which has afforded the author much pleasant mental recreation. Their consideration may not convey to other amateurs the pleasurable excitement of original pursuit; but all who are trying to unlock the mysteries of nature will find in them interesting suggestions; and it is hoped that they may excite inquiry which will lead to a substantial advance in scientific knowledge of the Constitution of Nature.

THE COSMIC LAW

OF

THERMAL REPULSION.

---

1. Preliminary

The immense projection of the tail of the great comet of 1882 led me to suspect that the phenomenon resulted from an outward push exerted by the radiant energy of the sun on the matter of the comet, and that the matter which thus yielded to the push and was projected outwards was that portion of the comet which had become superheated as the body approached the sun. The force causing the outward projection evidently came from the sun; the matter projected had been reduced to great tenuity; the form of the tail indicated that the outward push was exerted against the entire body of the comet, and that the particles projected yielded to the force as they became surcharged with the sun's radiant energy.

This explanation involved the hypothesis that the expansive force of heat was not confined to moving outwards the molecules of a separate mass of matter, as in the ordinary phenomenon of expansion, but that it was operative between the sun and bodies in space; in other words, that thermal energy exerted on all matter a push outwards from the centre of gravity, just as gravitation exerts a pull inwards towards the centre of gravity. Further reflection during subsequent years strengthened my belief in the truth of this hypothesis; and recent advances in physical science furnish evidence which appears to me to be sufficient, when considered in connection with other well-known physical phenomena, to prove the existence of the supposed Cosmic Law.

2. General Principles.

In attempting the induction of a Cosmic Law from the phenomena of nature, it is of course necessary to consider the whole subject of nature; and in doing so, the first thing which strikes the attention is the difference between those things in nature which are matter and those things which are not matter. For instance, the table on which I write, and the pen, ink, and paper with which I write, are matter; but the intelligence which directs the pen in making letters on the paper is not matter: it is force, imparting motion to matter.

All the things in nature which are subject to the law of gravitation are properly classified as matter; they have substance, weight, and occupy space. Those things which impart motion to matter are properly classified as forces. The principal forces in nature are : Intelligence, common to man and animals; the Vital Force, common to man, animals, and plants;

Cohesion, Chemical Affinity, Light, Electricity, and Magnetism, imparting motion to matter under varying conditions; and Gravitation and Heat, which are always present and operative where matter exists. In a general sense, the force exerted on molecules of matter by gravitation, cohesion, and chemical affinity is centripetal, or concentrative, while the force exerted by heat, and by light and electricity *when converted into heat*, is centrifugal, or expansive. But it is not necessary, for the purposes of this induction, to determine whether the foregoing classification is correct, nor to attempt to make any complete catalogue of all the forces of nature which have been the subject of scientific investigation.

We find, in nature, Time and Space, which are neither matter nor force, and it is possible that there are other things in nature, besides these two, which belong to the same classification; but in matter, force, and things which are neither matter nor force, we have the whole of nature. This term excludes, of course, the supernatural, and man's relation to it, and only embraces the entire domain of natural science. The chief object of this classification is to call attention to the fact that there is nothing in nature which is at the same time both matter and force. The inertia of matter being admitted, it is logically impossible for a thing to be matter and also force.

"Inertia is a purely negative though universal quality of matter: it is the property that matter cannot of itself change its own state of motion or rest. If a body is at rest, it remains so until some force acts upon it; if it is in motion, this motion can only be changed by the application of some force. This property of matter is what is expressed in Newton's first law of motion." (Ganot's Physics [Atkinson], art. 19.)

Motion is the transference of a thing from one part of space to another part; it is the opposite of rest. The forces of nature appear to possess inherent motion, just as matter possesses inherent inertia, but it is not necessary to determine this question. It is the motion which these forces impart to matter, and about which there is no question, that we are now to consider. Inertia is a universal quality of matter, and it is an essential quality of force that it overcomes inertia by imparting motion to matter whenever the conditions for its operation concur. It is this quality which justifies the classification of Intelligence with Gravitation. There is nothing else in common between the two things. Intelligence is not matter; but in one sense it is force; that is, when the conditions for its operation concur, it imparts motion to matter, and, in a division of nature into matter, force, and things which are neither matter nor force, Intelligence falls into the second division. The sole purpose of the classification is to emphasize the difference between matter and force. We have no knowledge of *force* disconnected from matter, and we have no knowledge of *matter* disconnected from force; but although united, they are separate and distinct elements in the constitution of nature, just as life and body are separate and distinct elements in the constitution of animal existence. When the animal dies the body part of it returns to the earth and air, from which it was formed, while the life disappears from observation entirely.

3. Operation of Natural Forces.

We have seen how the intelligence guides the pen in making the marks on the paper as I write; the bee, by force of an intelligence of an entirely different order, gathers honey from the flowers and transmits it to the hive. All the works of man, ancient and modern, constitute only a very small fraction of the matter to which motion has been imparted by the force of intelligence. It is measured only by the aggregate activity of every living being which has existed from the beginning of the world.

Every molecule constituting the body of a living thing, animal or vegetable, now on the earth or that has ever existed, has been put in motion by the vital force. It is this wonderful thing which has eliminated from the storehouse of nature the molecules suitable for my finger-nails, for the hairs of my head, for the muscles, nerves, brain, and other organs of my body, and carried each to its proper place in the living organism. And this has been done in respect to every molecule constituting a part of the body of every animal and of every plant now living or that ever has lived on the earth. The quantity of matter, every molecule of which has been moved by the vital force, is simply incalculable. In many cases the same molecule has been moved probably many millions of times, as it became successively a part of the body of different animals and plants.

The operation of the force of gravitation is well known. We see it in the mighty rivers running to the sea, and in the motions of the planetary bodies, and in many other forms. It operates between separate bodies of matter, as well as between the molecules constituting a single body. Newcomb tersely defines it as follows: "Every particle of matter in the universe attracts every other particle with a force directly as their masses, and inversely as the square of the distance which separates them." (Pop. Astr. [6th Ed.], p. 81.)

The force operates on the molecules constituting a separate body of matter, imparting motion to them in the same way that it imparts motion to the separate bodies themselves.

This molecular motion imparted by gravitation to matter has not received scientific consideration commensurate with its importance. It is this force which causes a drop of water or a drop of molten lead to assume the form of a sphere. Each molecule of the drop is drawn by this force of gravitation to the centre of gravity of the drop, and a perfect sphere is the necessary result of this struggle of each molecule to reach the centre of gravity, because in that form the average distance of all the molecules from the centre of gravity is less than in any other form. A particle of unmelted lead will not assume the form of a sphere, because the force of cohesion counteracts the gravitational attraction of the molecules for each other. In the process of cooling while falling from the top to the bottom of a shot-tower, the force of cohesion again becomes dominant on the molecules constituting the drop of molten lead, and if the process has been unobstructed, the drop remains a mathematically perfect sphere.

But if the drop of water or of molten lead, before the force of cohesion has regained control, reaches the earth, the drop spreads out flat, because the force of the earth's gravitational attraction on each of the molecules constituting the drop is greater than the force of the attraction of the molecules for each other. If the earth's attraction of gravitation on the molecules constituting water could be removed, leaving its attraction for the body of water, and leaving the molecules free to obey their attraction for each other, every body of water on the earth's surface would be instantly converted into a globule, perfectly round and smooth, resting like a huge glass ball in the bed of the lake, river, or sea which now holds the water. And if the force of cohesion could be removed from solid matter, as it is from lead when it is melted, the whole surface of the earth would become at once absolutely smooth and perfectly level ; life would be impossible, and the whole economy of nature would be reversed.

We see, therefore, the importance of the earth's attraction of gravitation on the molecules constituting the body of a liquid, as well as the importance of the same attraction upon the body itself, and the overwhelming necessity for the force of cohesion to counteract the motion which, but for this force, would be imparted by gravitation to the molecules constituting solid matter.

The extent to which, in its natural state, matter is subjected to the force of chemical affinity is shown by the fact that the elementary substances are rarely, if ever, found in simple form, and a chemical reaction is necessary to unlock them from the embrace into which they have been driven by the force of chemical affinity.

The motion imparted to matter by heat is expansion; the molecules are pushed asunder, and the extent of this motion is proportional (with certain well-known exceptions) to the amount of heat absorbed by the body in which the expansion occurs. What we wish to demonstrate now is, that this same force of heat pushes asunder masses of matter as well as molecules; and that therefore a mass of matter surcharged with the sun's radiant energy has thereby imparted to it the centrifugal motion possessed by the radiant energy before its absorption.

4. An Illustration.

A running stream of clear water containing fish, and a boy fishing for them, furnish a beautiful illustration of the principal forces of nature in their ordinary conjunctive operation. The water runs in obedience to the force of gravitation drawing it from the highlands, where it fell as rain, to the level of the sea. If this force was suspended, the water would cease to flow. If this force of the earth's attraction of gravitation on the mass of water remained in full operation, but its attraction on the separate molecules constituting the water was suspended, leaving the molecules subject to their attraction for each other, the water would instantly cease to flow, and would become globules rolling down the bed of what had been the stream. Thus in the stream itself we see the action of gravitation on both the mass and the molecules of water. It is the force of cohesion which holds together the solid banks in which the stream is confined. But for this force, the earth's attraction of gravitation would cause the solid surface to flatten out as smooth as a sheet of ice, and the water would spread out like a drop on the surface of a glass ball. In the water itself we have an example of the action of chemical affinity not surpassed by anything in nature's vast workshop. Oxygen and hydrogen, two transparent invisible gases, at a temperature far above that ordinarily existing on the earth's surface, unite, and the product of their union is aqueous vapor, also a transparent invisible gas ; but when its temperature, sinking towards terrestrial normals, reaches 100° C. (212° F.), this gaseous product of a gaseous combination has become the liquid constituting the stream.

But it is the force of thermal energy which holds apart the molecules of this water, and maintains its liquid form. If this force were withdrawn, the rippling stream would be converted instantly into a mass of solid ice. And light, another form of the same force, enables us to see what is going on.

The grass, weeds, and trees growing on the banks of the stream show how the vital force, aided by other forces, has changed inert matter into the varying forms which they present; and a dead leaf floating on the surface shows the immense difference which this force produces in nature—the difference between life and death. It is this vital force also which enables the fish to swim up stream, thus overcoming the grav-

itational motion of the water, while a dead fish drifts
with the current like the dead leaf.

A fish sees a fly drop on the water, and immediately
darts after it and swallows it as food. It is not the
mere vital force which exists in trees and grass, as
well as in the fish, which causes this motion. It is
the force of intelligence—the capacity to apprehend—
which causes the fish to know its food and to seize it.
The fish had no agency in causing the fly to drop on
the water; but the boy, put in motion by the force of
an intelligence of a much higher order, catches another
fly, sticks a hook through it, and throws it on the
water: the intelligence of the fish again puts him in
motion towards this second fly, as towards the first,
but this time the force of the fish's intelligence is supplemented by the force of the boy's intelligence, and
the resultant motion lands the fish on dry ground.
It is thus obvious that gravitation, cohesion, chemical
affinity, thermal energy, the vital force, and the force
of intelligence are all present, in effect or in operation,
whenever a boy catches a fish in a running brook.
All the forces of nature are as omnipresent as gravitation, and never fail to appear where the conditions for
their operation concur.

5. Field of Operation of Natural Forces.

Intelligence without life is impossible; and the vital
force without gravitation, cohesion, chemical affinity,
and thermal energy cannot exist: cohesion, chemical
affinity, electricity, and the magnetic force are doubtless always present where matter exists, but are oper-

ative only under certain conditions; while gravitation and heat are not only always present, but are always active. We have seen that gravitation is universal, and Tyndall says: "There is no body in nature absolutely cold; and every body not absolutely cold emits rays of heat." (Fragments of Science, p. 28.) Matter is inert, but there is no place or condition in nature where it can escape for a moment entirely from either one of these two ever-present forces.

6. Forms of Matter.

Matter, subject to the operation of these forces, is in three forms—solid, liquid, and gaseous; and whether it is in the one or the other of these forms seems to depend on the intensity of its subjection to the one or the other of these forces. There is apparently something in the constitution of matter which determines the measure of its respective subjection or resistance to these forces, and, consequently, the form of the matter. These forces of nature do not explain why, at ordinary terrestrial temperatures, iron is a solid, mercury a liquid, and hydrogen a gas; nor why it requires an immense force to solidify hydrogen, a less force in the same direction to solidify mercury, and a force probably as great, but in the opposite direction, to vaporize iron. But whatever may be the cause of this difference in the form of matter, the difference itself is subordinate to the forces of nature. It has been long known that many substances, notably water, pass readily from one form to another, from liquid to solid ice, then back to liquid, and then to the gaseous form of steam. But recent experiments have demonstrated that no matter is in a permanent form; all solids can be vaporized and all gases liquefied, and, doubtless, solidified also, by a proper application of the forces of nature. When a solid is converted into a liquid, it has all the qualities of an original or normal liquid; and when a liquid is completely converted into vapor, it has all the qualities of a gas. Before the transformation is complete,—that is, before all the liquid has been converted into vapor,—it is called a *saturated* vapor; and in this condition of partial transformation,—as, for instance, steam in a boiler still containing water,—the vapor has not all the qualities of gas: but when the transformation has been completed by vaporizing all the liquid, the vapor is then said to be unsaturated, and has all the qualities of other gases.

" Unsaturated vapors behave in all respects like gases. And it is natural to suppose that what are ordinarily called *permanent gases* are really unsaturated vapors. For the gaseous form is accidental, and is not inherent in the nature of the substance. At ordinary temperatures sulphurous anhydride is a gas, while in countries near the poles it is a liquid; in temperate climates ether is a liquid,—at a tropical heat it is a gas. And just as unsaturated vapors may be brought to the state of saturation, and then liquefied by suitably diminishing the temperature or increasing the pressure, so by the same means gases may be liquefied. But as they are mostly very far removed from this state of saturation, great cold and pressure are required. Some of them may indeed be liquefied either by cold or pressure; for the majority, however, both agencies must be simultaneously employed. The recent researches of Cailletet and of Pictet have shown that the distinction *permanent* gas no longer exists, now that all are liquefied." (Ganot's Physics [Atkinson], art. 380.)

See also "Heat a Mode of Motion," by Tyndall, pp. 142 to 146.

It is important to bear this in mind, because it demonstrates that matter does not lose its identity by any change of form into which it may be forced. Iron remains iron, when liquefied and vaporized; and when the force is removed, it comes back to the liquid, and then to the solid iron, from which it started. Water loses none of its constituent elements by being converted into ice in one direction or into steam in the opposite direction; and when its normal liquid form is restored, by removing the force, it shows nothing but more or less purification as a result of the change of form. This is true of other substances: the application of force which changes the form, does not necessarily or ordinarily affect the identity of the substance itself.

7. The Earth's Attraction on Liquids and Gases.

All liquids and gases within the reach of man's experiments are subjected to the force of the earth's attraction of gravitation to an extent that entirely changes the shape of the body or mass from what it would be if this force was removed. We have seen how this force flattens out a drop of water or of melted lead, which, if left to the force of gravitation operating on its own molecules, would remain a perfect sphere. This force operates with even more effect on the molecules of a gas than on the molecules of a liquid. It is this force which not only spreads out the atmosphere flat all around the surface of the earth, as it does water, but, by reason of the greater compressibility of the atmosphere, this attractive force causes an immense difference in the density of the atmosphere at the earth's surface and at the confines of interstellar space.

The attraction of gravitation is mutual; and the fact that the earth attracts the separate molecules of liquids and gases is a demonstration that each of these molecules separately possesses the power of gravitational attraction,—in other words, that every separate molecule, as well as every mass or aggregation of molecules, is subject to the law of gravitation, and exerts gravitational force on every other molecule.

8. Effect of Gravitation on Molecules of Gas.

This demonstrates that the molecules of gas are pulled inwards by their mutual gravitational attraction; and, as they are necessarily subject to the law of inertia, nothing could be more absurd than the hypothesis that, if the molecules of gas "were in space, where no external force could act upon them, they would fly apart, and disappear in immensity."

There can be no more necessity for an external force to prevent the molecules of gas from flying apart and disappearing in space than there is for such a force to prevent the planets from doing the same thing. The hypothesis is not only in direct contradiction of the observed phenomena of the force of gravitation operating on molecules of the atmosphere, but it involves a denial of the law of inertia. This law, like the law of gravitation, binds every separate molecule of hydrogen gas as effectually as it does the planet Jupiter. Molecules of matter, in solid, liquid, or gaseous form, cannot fly apart or together except in obedience to an impelling force. Physical science has not yet furnished any accurate description of a molecule ; and it is not necessary, for the purposes of this inquiry, to determine whether molecules are solid bodies held apart by an intervening ether, which is increased in expansion and forced out in contraction ; or whether they are elastic bodies, which occupy less space when they are contracted than when expanded ; or whether they are in constant vibratory motion, according to the kinetic theory. What we do know certainly, is that they are particles of matter, and that each one of them separately is subject to the law of gravitation and to the law of inertia ; and that the application of heat causes an expansion either by increasing the size of the molecules themselves, or by increasing the spaces which separate them, or by increasing the vibratory motion.

9. EFFECT OF HEAT ON MATTER.

Where the heating is uniform, the expansion is the same in all directions ; and this demonstrates that the direction of the force is from the centre of gravity outwards,—but the consequent motion is in the direction of least resistance. If a bar of railroad iron, placed with one end against a wall, is heated by the sun, the elongation is at the other end. In contracting, the direction of the force is towards the centre of gravity, which is in the middle of the elongated bar; consequently, when the bar cools, a space will be found between the wall and the end of the bar. Tyndall describes graphically the motion imparted by heat to molecules. He says: "As already indicated, the atoms or molecules, thus vibrating, and ever, as it were, seeking wider room, urge each other apart, and thus cause the body of which they are constituents to expand in volume. By the force of cohesion, then, the molecules are held together; by the force of heat they are pushed asunder: and on the relation of these two antagonistic powers, the aggregation of the body depends. Every fresh increment of heat pushes the molecules more widely apart; but the force of cohesion, like all other known forces, acts more and more feebly as the distance through which it acts is augmented. As, therefore, the heat grows strong, its opponent grows weak, until finally the particles are so far loosened from the thrall of cohesion as to be at liberty, not only to vibrate to and fro across a fixed position, but also to roll or glide around each other. Cohesion is not yet destroyed, but it is so far modified that the particles, while still offering resistance to being torn directly asunder, have their lateral mobility over each other's surfaces secured. *This is the liquid condition of matter.*

"In the interior of a mass of liquid, the motion of every molecule is controlled by the molecules which surround it. But when we develop heat of sufficient power, even within the body of a liquid, the molecules break the last fetters of cohesion, and fly asunder to form bubbles of vapor. If, moreover, one of the surfaces of the liquid be quite free,—that is to say, uncontrolled either by a liquid or a solid,—it is easy to conceive that some of the vibrating superficial molecules will be jerked entirely away from the liquid, and will fly with a certain velocity through space. *Thus freed from the influence of cohesion, we have matter in the vaporous or gaseous form.*" (Heat a Mode of Motion, p. 116.)

His statement that the molecules are "vibrating" is mere theory, and, whether true or false, cannot be used as a term in legitimate induction. He ignores the fact that the molecules are held together by gravitation as well as by cohesion, and that the freed molecules flying through space are each subject to the law of gravitation and also to the law of inertia, and that their velocity results entirely from the heat-force imparted or applied to them. With these additions, his description would be both accurate and graphic.

Recurring to the subject again, he says (pp. 190, 191): "We have hitherto confined our attention to the heat consumed in the molecular changes of solid and liquid bodies while they continue solid or liquid. We shall now direct our attention to the phenomena which accompany changes of the state of aggregation. When sufficiently heated, most solids become liquids; and when still further heated, the liquids become gases. Let us consider the case of ice, and follow it through the entire cycle of its changes. Take, then, a block of ice at a temperature of 10° C. below zero. The ice being gradually warmed, a thermometer fixed in it rises to 0°. At this point the ice begins to melt, and the thermometric column, which rose previously, becomes stationary. The warmth is still applied, but no augmentation of temperature is shown by the thermometer; and not until the last film of ice has been removed from the bulb does the mercury resume its motion. It then ascends again, reaching in succession 30°, 60°, 100°. Here steam-bubbles appear in the liquid, it boils, and from this point onwards the thermometer remains stationary at 100°.

"To simply liquefy a pound of ice, as much heat is expended as would raise a pound of water 79°.4 C., or 79.4 pounds of water 1°, in temperature; while to convert a pound of water at 100° C. into a pound of steam of the same temperature, 537.2 times as much heat is required as would raise a pound of water one degree in temperature. The former number, 79°.4 C. (or 143° F.), represents what has been hitherto called the latent heat of water; and the latter number, 537°.2 C. (or 967° F.), represents the latent heat of steam. It was manifest to those who first used these terms that, throughout the entire time of melting, and throughout the entire time of boiling, heat was communicated; but inasmuch as this heat was not revealed by the thermometer, it was said to be rendered latent. The fluid of heat was supposed to hide itself, in some unknown way, in the interstitial spaces of the water and of the steam. According to our present theory, the heat expended in melting is consumed in overcoming molecular attractions, and in conferring potential energy upon the separated molecules or their poles. It is virtually the lifting of a weight. So likewise, as regards vaporizing, the heat is consumed in separating the molecules still farther asunder, and in conferring upon them a weight of potential energy. When the heat is withdrawn, the vapor condenses; that is to say, the molecules fall together with an energy equal to that employed to separate them. By further chilling, the water crystallizes to ice, restitution being made in both cases of the precise quantity of heat consumed in the acts of fusion and vaporization.

" The act of liquefaction consists almost solely of interior work—of work expended in moving the atoms into new positions. The act of vaporization is also, for the most part, interior work; to which, however, must be added the exterior work of forcing back the atmosphere when the liquid becomes vapor."

This beautiful description is also defective in failing to notice that it is the force of gravitation which brings the molecules together when the heat is withdrawn and the vapor condenses. The force of cohesion is entirely overcome when the liquid becomes vapor; but the force of gravitation, operating between each separate molecule and every other molecule, remains,—diminished, it is true, by the square of the increased distance between them, but still operative. The moment the outward push of the heat begins to diminish, the inward pull of gravity becomes dominant; and the molecules, moved outwards from the centre of gravity by the heat, begin to move inwards towards the centre of gravity under the force of gravitation.

This outward push of heat and inward pull of gravity are more obvious to observation in solids than in liquids or gases, but they exist in all forms of matter; the outward push being the result of the expanding force of thermal energy, and the inward pull being the result of the law of gravitation, as determined and demonstrated by Newton. The phenomenon of condensation itself demonstrates the presence and operation of the law of gravitation on the molecules of the vapor; for, the moment the process of condensation has so far progressed as to be the subject of observation, the aggregated molecules are found grouped into drops, which, under the force of gravitation operating on the molecules, are spherical.

10. Conjugal Antagonism of Heat and Gravity.

It is obvious, therefore, that on a body of terrestrial matter the effect of the force of gravitation is to impart motion to the molecules inwards, the direction of the force being towards the centre of gravity of the body; and that the effect of the force of heat is to impart motion to the molecules outwards from the centre. The force of gravitation is constant; and the force of heat is constant also, no place or condition having been discovered where there was matter but no heat. An absolute zero of temperature, where matter exists, is no more possible than the existence of matter without gravitation; for heat is obviously as universal and omnipresent in nature as gravitation itself. It exists in space wherever the sun and stars shine, and in the middle of the earth, and doubtless in the middle of the other planets also it exists in immeasurable quantity. But the quantity of heat received by terrestrial matter is variable; and when such a body receives an increase, the circumstances under which this heat will be lost, are also variable. In fact, all inert matter near the earth's surface is being heated or cooled, practically, all the time. Whenever the heat in any body diminishes, there is a contraction; that is, the molecules composing the body are put in motion towards the centre of gravity: and whenever the heat increases, there is an expansion; that is, the molecules are put in motion outwards from the centre of gravity. There are some well-known exceptions in both processes. Under peculiar circumstances cold causes matter to expand (as in the formation of ice), and in others heat causes a contraction; but these exceptions will doubtless be traced to the operation of forces not inconsistent with the general rule.

In all terrestrial matter, solid, liquid, or gaseous, in which heat and cold produce expansion and contraction, the conjugal antagonism of the opposing forces on the molecules composing the matter is constant. Precisely what relation the force of cohesion and the force of chemical affinity bear to this struggle is not yet known. We do know that, when heat so far prevails as to produce vaporization, cohesion disappears; and at temperatures varying with different substances, the force of chemical affinity disappears also, and the matter returns to its original elements. From the heat required to reduce ice to water and water to vapor, and to reduce certain chemical combinations, it appears probable that both cohesion and chemical affinity are, in a certain sense, adjuncts of gravitation; but, as before stated, it is not necessary to this induction to determine more definitely the operation of these forces in the economy of nature. Heat and gravitation are the two forces, neither of which is ever absent from matter under any circumstances, and we have seen that they respectively impart motion to the molecules of terrestrial matter in opposite directions.

11. Gravitation and Thermal Energy on Masses of Matter.

It is manifest that gravitation exerts the same force and imparts the same motion to masses that it does to the molecules of matter, and that it is this force which prevents centrifugal motion in planetary bodies; and what we propose to show is, that Thermal Energy also exerts the same force and imparts the same motion to masses that it does to molecules of mattter, and that it is this force which prevents centripetal motion in planetary bodies.

It would be an anomaly in nature if the conjugal antagonism between these forces, in imparting motion to molecules of matter, did not exist also in respect to masses of matter, which are but aggregations of molecules. But we are not left to conjecture on this subject: the telescope and spectrum analysis have disclosed facts which seem to demonstrate that the radiant energy of the sun imparts to planetary matter an outward push similar to that imparted to molecules of a body of terrestrial matter by heat.

12. PLANETARY MATTER—COMETS.

Newton demonstrated that the force of gravitation was interplanetary, and that through this force the sun controlled the orbital motion of the planets. This force, thus passing through interstellar space from the sun to the planets, from planet to planet, and from each back to the sun, is identically the same force which causes a globular drop of water to spread out flat when it reaches the earth's surface. It has been known from the dawn of intelligence that the radiant energy of the sun also crossed this vacuum space and warmed the earth ; but recent observations of comets have demonstrated that, in respect to that class of planetary bodies, the radiant energy of the sun imparted to the matter of the comet, or to some part of it, an outward push from the sun apparently identical with that which heat imparts to the molecules of terrestrial matter.

These bodies, like all other matter, are subject to the law of gravitation ; and their extreme tenuity and want of cohesion have enabled astronomers to observe the direct effect of the sun's outward push.

Professor Young states the result of these observations as follows : "The orbits of these bodies are now thoroughly understood, and their motions are calculated with as much accuracy as the nature of the observations permits ; but we find in their physical constitution some of the most perplexing and baffling problems in the whole range of astronomy—apparent paradoxes which as yet have received no satisfactory explanation.   While comets are evidently subject to gravitational attraction, as shown by their orbits, they also exhibit evidence of being acted upon by powerful repulsive forces emanating from the sun." (Young's General Ast., p. 403.)

It has been demonstrated by spectrum analysis that the tail of a comet is a projection of cometary matter. Miss Clerke says : " The mystery of comets' tails has been to some extent penetrated—so far at least that, by making certain assumptions strongly recommended by the facts of the case, their forms can, with very approximate precision, be calculated beforehand.   We have, then, the assurance that these extraordinary appendages are composed of no ethereal or supersensual stuff, but of matter such as we know it, and subject to the ordinary laws of motion, though in a state of extreme tenuity.   This is unquestionably one of the most remarkable discoveries of our time." (Hist. of Ast. during the Nineteenth Cent., p. 392.)

Professor Newcomb has this to say about the phenomena of solar repulsion : " Altogether a good idea of the operations going on in a comet will be obtained if we conceive the nucleus to be composed of water or other volatile fluid which is boiling away under the heat of the sun, while the tail is a column of steam rising from it. . . . We now meet a question to which science has not yet been able to return a satisfactory answer.   Why does this mass of vapor always fly away from the sun ?   That the matter of the comet should be vaporized by the sun's rays, and that the nucleus should thus be enveloped in a cloud of vapor, is perfectly natural, and entirely in accord with the properties of matter which we observe around us. But, according to all known laws of matter, this vapor should remain around the head, except that the outer portions would be gradually detached and thrown off in separate orbits. There is no known tendency of vapor, as seen on the earth, to recede from the sun, and no reason why it should so recede in the celestial spaces. Various theories have been propounded to account for it; but as they do not rest on causes which we have verified in other cases, they must be regarded as purely hypothetical." (Popular Ast. [Newcomb], p. 414.)

Miss Clerke, in her most excellent "History of Astronomy," has collected all the learning on the subject. She says: "The ingenious view recently put forward by M. Bredichin, of Moscow, as to the connection between the form of these appendages and the kind of matter composing them was very clearly anticipated by Olbers. The amount of tail curvature, he pointed out, depends in each case upon the proportion borne by the velocity of the ascending particles to that of the comet in its orbit: the swifter the outrush, the straighter the resulting tail. But the velocity of the ascending particles varies with the energy of their repulsion by the sun, and this again, it may be presumed, with their quality. Thus multiple tails are developed when the same comet throws off, as it approaches perihelion, specifically distinct substances. The long, straight ray which proceeded from the comet of 1807, for example, was doubtless made up of particles subject to a much more vigorous solar repulsion than those formed into the shorter, curved emanation issuing from it nearly in the same direction. In the comet of 1811 he calculated that the particles expelled from the head travelled to the remote extremity of the tail in eleven minutes, indicating by this enormous rapidity of movement [comparable to that of the transmission of light] the action of a force greatly more powerful than the opposing one of gravity." (Hist. of Ast., pp. 125, 126.) Describing Donati's comet, she says (p. 370): "At Pulkowa, on the 16th of September, Winnecke observed a faint outer envelope, resembling a veil of almost evanescent texture, flung somewhat widely over the head. Next evening, the first of the 'secondary' tails appeared, possibly as part of the same phenomenon. This was a narrow, straight ray forming a tangent to the strong curve of the primary tail, and reaching to a still greater distance from the nucleus. It continued faintly visible for about three weeks, during the part of which time it was seen in duplicate; for from the chief train itself, at a point where its curvature abruptly changed, issued, as if through the rejection of part of its materials, a second beam nearly parallel to the first, the rigid line of which contrasted singularly with the softly diffused and waving aspect of the plume of the light from which it sprang. Olbers's theory of unequal repulsive forces was never more beautifully illustrated. The triple tail was a visible solar analysis of cometary matter."

Her summary of the learning on the subject of cometary tail formation is very clear: "It is now nearly a quarter of a century since M. Bredichin, director of the Moscow Observatory, directed his attention to these curious phenomena. His persistent inquiries on the subject, however, date from the appearance of Coggia's comet in 1874. On computing the value of the repulsive force exerted in the formation of its tail, and comparing it with the values of the same force arrived at by him in 1862 for some other conspicuous comets, it struck him that the numbers representing them fell into three well-defined classes. 'I suspect,' he wrote in 1877, 'that comets are divisible into groups, for each of which the repulsive force is the same.' This idea was confirmed on fuller investigation. In 1882 the appendages of 36 well-observed comets had been reconstructed theoretically, without a single exception being met with to the rule of the three types. A further study of 40 comets led, however, in 1885, to a modification of the numerical results previously arrived at.

"In the first of these, the repellent energy of the sun is fourteen times as strong as his attractive energy; the particles forming the enormously long, straight rays projected outward from this kind of comet leave the nucleus with a mean velocity of just seven kilometres per second, which, becoming constantly accelerated, carries them in a few days to the limits of visibility. The great comets of 1811, 1843, and 1861, that of 1744 (so far as its principal tail was concerned), and Halley's comet at its various apparitions, belonged to this class. For the second type the value of the repulsive force employed is less narrowly limited. For the axis of the tail it exceeds by one and one-

tenth (1.1) the power of solar gravity; for the ante-
rior edge it is more than twice (2.2), for the posterior
only half, as strong.   The corresponding initial velocity (for the axis) is 1500 metres a second, and the resulting appendage a cimeter-like or plumy tail, such
as Donati's and Coggia's comets furnished splendid
examples of.   Tails of the third type are constructed
with forces of repulsion from the sun varying from
one-tenth to three-tenths that of his gravity, producing
an accelerated movement of attenuated matter from the
nucleus, beginning at the leisurely rate of 300 to 600
metres a second.   They are short, strongly bent, brush-
like emanations, and, in bright comets, seem only to
be found in combination with tails of the higher classes.
Multiple tails, indeed,—that is, tails of different types
emitted simultaneously by one comet,—are perceived,
as experience advances and observation becomes closer,
to be rather the rule than the exception.

"Now, what is the meaning of these three types? Is
any translation of them into physical fact possible?
To this question Bredichin supplied in 1879 a plausible answer.   It was already a current surmise that
multiple tails are composed of different kinds of matter, differently acted on by the sun.   Both Olbers and
Bessel had suggested this explanation of the straight
and curved emanations from the comet of 1807; Norton had applied it to the faint light-tracks proceeding
from that of Donati's; Winnecke, to the varying devi-
ations of its more brilliant plumage.   Bredichin defined and ratified the conjecture.   He undertook to
determine (provisionally as yet) the several kinds of
matter appropriated severally to the three classes of

tails. These he found to be hydrogen for the first, hydro-carbons for the second, and iron for the third. The ground for this apportionment is that the atomic weights of these substances bear to each other the same inverse proportion as the repulsive forces employed in producing the appendages they are supposed to form ; and Zöllner had pointed out in 1875 that the 'heliofugal' power by which comets' tails are developed would in fact be effective just in that ratio. Hydrogen, as the lightest known element,—that is, the least under the influence of gravity,—was naturally selected as that which yielded most readily to the counter-persuasions of electricity. Hydro-carbons had been shown by the spectroscope to be present in comets, and were fitted by their specific weight, as compared with that of hydrogen, to form tails of the second type ; while the atoms of iron were just heavy enough to compose those of the third, and, from the plentifulness of their presence in meteorites, might be presumed to enter, in no inconsiderable proportion, into the mass of comets. These three substances, however, were by no means supposed to be the sole constituents of the appendages in question. On the contrary, the great breadth of what for the present were taken to be characteristically 'iron' tails, was attributed to the presence of many kinds of matter of high and slightly different specific weights ; while the expanded plume of Donati was shown to be in reality a whole system of tails, made up of many substances, each spreading into a hollow cone, more or less deviating from, and partially superposed upon, the others.

" Never was a theory more promptly or profusely illustrated than this of Bredichin." (History of Astronomy of 19th Century [Clerke], pp. 393-6.)

These authorities, without going further into the details of the observations, establish the proposition that the radiant energy of the sun imparts centrifugal motion to parts of the matter of the comets—to some parts much more than to the others. The fair inference from these facts is that the sun's radiant energy imparts an outward push to the entire mass of the comet, the push in respect to the different parts of the comet being in proportion to the amount of heat absorbed by these different parts respectively. This is certainly more probable than to suppose that a new force originates or is developed in the comet itself, and that this new force causes the projecting away from the sun. The appearance is that the sun's radiant energy operates like a wind and blows away the loosened particles. The force of cohesion existing to a very limited extent, the comet must be held together by the attraction of gravitation between the molecules ; and consequently it appears that the cometary matter most affected by the attraction of gravitation presents the greatest resistance to the repellent force.

It is manifest, from the effect of gravitation and expansion on terrestrial matter, that a gaseous body in space, subjected to no force except those two, would undergo a constant change of form as one or the other force predominated.

It is not known to what extent expansion of terrestrial matter is the result of the electrical form of thermal energy. Nor is it known to what extent the "heliofugal" power exerted by the sun on cometary matter, is the result of this form of thermal energy. We know that the heliofugal power is transmitted by the sun's rays from the sun across the intervening vacuum to the comet, and that the centrifugal force of expansion is imparted to the molecules of terrestrial matter by the application of heat to the mass; but precisely how the force is imparted in either case is yet a mystery.

13. MOTION IMPARTED BY HELIOFUGAL POWER RESISTED BY COHESION AND GRAVITATION.

Pretermitting these questions, however, as not necessary to the present determination, it will be seen that all the observed phenomena of cometary transformations are not only consistent with but apparently result from the inference above stated, namely, that the radiant energy of the sun imparts to cometary matter centrifugal force—the amount of force imparted being in proportion to the heat absorbed, and the resultant motion being in the direction of the least resistance. Whenever any force is applied to matter, the resultant motion is in the direction of the force, unless there is less resistance in some other direction, and then the resultant motion is in the direction of least resistance. It is important to bear this law in mind in considering the motion imparted by heliofugal power to cometary matter.

Comets are bodies small in mass but large in volume. They appear to have no separate axial motion, but revolve around an axis of the sun, just as the moon revolves around an axis of the earth ; and consequently, the same face is always presented to the sun, and the entire force of the sun's energy is received in the same place at all times. The intensity of the force of radiant energy is inversely as the square of the distance, and is in proportion to the angle of incidence. The side or end of the comet next to the sun is not only exposed to the direct action of the sun's radiant energy, while the outer side receives only such heat as passes entirely through the comet, but this comparatively small quantity is diminished by the square of the distance from the inner to the outer parts of the comet. It follows that the part of the comet next to the sun is heated much more intensely than the outside, and that the relative difference in the heating of the inner and the outer parts of the comet increases as the comet approaches the sun.

If the heliofugal power is imparted in proportion to the amount of heat absorbed, it is manifest that, as compared with the other parts of the comet, an immense outward push is imparted to the portion of the comet next to the sun. The direction of this force is outward from the sun ; the resistance to the resultant motion is from the cohesion of the less heated parts of the comet, and the force of gravitation on the matter to which outward push is imparted.

We have already seen that astronomers ascribe the difference in the curvature of the comet's tail to the effect of gravitation on the matter composing the tail ; the tails straight, or nearly so, being composed of hydrogen, those curved slightly being composed of hydro-carbons, and those most curved being composed of the vapors of iron and other heavy substances. There are other phenomena, which appear to result from the resistance to projection presented by the cohesion of the less heated parts of the comet.

Professor Young thus describes the constituent parts of a comet : " (a) The essential part of a comet —that which is always present and gives it its name— is the coma or nebulosity, a hazy cloud of faintly shining matter, which is usually nearly spherical or oval in shape, though not always so.

" (b) Next we have the nucleus, which, however, is not found in all comets, but commonly makes its appearance as the comet approaches the sun. It is a bright, more or less starlike point near the centre of the coma, and is the object usually pointed on in determing the comet's place by observation. In some cases the nucleus is double or even multiple; that is, instead of a single nucleus there may be two or more near the centre of a comet. Perhaps three comets out of four present a nucleus during some portion of their visibility.

" (c) The tail or train is a streamer of light which ordinarily accompanies a bright comet, and is often found even in connection with a telescopic comet. As the comet approaches the sun, the tail follows it much as the smoke and steam from the locomotive trail after it. But that the tail does not really consist of matter simply left behind in that way is obvious from the fact that as the comet recedes from the sun, the train precedes it instead of following. It is always directed away from the sun, though its precise position and form is to some extent determined by the

comet's motion. There is abundant evidence that it
is a material substance in an exceedingly tenuous condition, which in some way is driven off from the comet
and then repelled by some solar action. (See also
art. 736).

" (*d*) *Envelopes and Jets.*—In the case of a very
brilliant comet, its head is often veined by short jets
of light which appear to be continually emitted by the
nucleus ; and sometimes instead of jets the nucleus
throws off a series of concentric envelopes, like hollow
shells, one within the other. These phenomena, however, are not usually observed in telescopic comets to
any marked extent." (Young's Gen. Astr., art. 714.)

The same writer speaks of the development of jets
and envelopes and the formation of the tail as follows : " When a comet is first seen at a great distance
from the sun, it is ordinarily a mere roundish, hazy
patch of faint nebulosity, a little brighter near the
centre.

" As the comet draws near the sun, it brightens, and
the central condensation becomes more conspicuous
and sharply defined, or starlike. Then, on the side
next the sun, the newly formed nucleus begins to emit
jets and streamers of light, or to throw off more or
less symmetrical envelopes, which follow each other
concentrically at intervals of some hours, expanding
and growing fainter as they ascend, until they are
lost in the general nebulosity which forms the head.
During these processes the nucleus continually
changes in brilliancy and magnitude, usually growing
smaller and brighter just before the liberation of each
envelope. When the jets are thrown off, the nucleus

seems to oscillate, moving slightly from side to side; but no evidences of a continual rotation have ever been discovered." (*Idem,* art. 727.)

" *Formation of the Tail.*—The material which is projected from the nucleus of the comet, as if repelled by it, is also repelled by the sun, and driven backward, still luminous, to form the train. (At least this is the appearance.) The researches of Bessel, Norton, and especially the late investigations of the Russian Bredichin, have shown that this theory—that the tail is composed of matter repelled by both the comet and the sun—not only accounts for the phenomena in a general way, but for almost all the details, and agrees mathematically with the observed position and magnitude of the tail on different dates. (*Idem,* art. 728.)

" *Anomalous Tails and Streamers.*—It is not very unusual for comets to show tails of two different types at the same time, as, for instance, Donati's comet. But occasionally stranger things happen, and the comet of 1757 is reported to have had six tails, diverging like a fan. Winnecke's comet of 1877 threw out a tail laterally, making an angle of about 60 degrees with the normal tail, and having the same length—about one degree. Pechüle's comet of 1880 (a small one), besides the normal tail, had another of about the same dimensions erected towards the sun; streamers of considerable length so directed are not very infrequent. The great comet of 1882 presented a number of peculiarities, which will be mentioned in the more particular description of that body which is to follow. Most of these anomalies are as yet entirely unexplained." (*Idem*, art. 736.)

All these phenomena of jets, envelopes, streamers, multiple and anomalous tails may well result from the resistance to direct outward motion presented by the unheated parts of the comet, thus forcing the projected particles into directions of least resistance. A rebound directly towards the sun, as was observed in the comet of 1882, would result if the particles met in any part of the body of the comet a sufficiently stable resistance to outward motion.

14. DIFFERENCE IN MASS AND DENSITY OF COMETS.

It is obvious from astronomical observations that comets differ in density, and it is probable that the same comet differs very widely in density in different parts of its orbit, being gaseous, or nearly so, at perihelion, and solid, or nearly so, at aphelion. But it is more than probable that there is an appreciable difference in both the mass and density of the same comet at each approach to perihelion. These bodies have a wonderful power of gathering themselves together after being badly scattered, as shown by Miss Clerke, in her description of the transformation of Encke's comet. She says: "The fact that comets contract in approaching the sun had been noticed by Helius; Pingrè admitted it with hesitating perplexity; the example of Encke's comet rendered it conspicuous and undeniable. On the 28th of October, 1828, the diameter of the nebulous matter composing this body was estimated at 312,000 miles. It was then about one

and a half times as remote from the sun as the earth
is at the time of the equinox. On the 24th of December following, its distance being reduced by nearly
two-thirds, it was found to be only 14,000 miles across.
That is to say, it had shrunk, during those two months
of approach, to $\frac{1}{11000}$ part of its original volume. Yet
it had still seventeen days' journey to make before
reaching perihelion. The same curious circumstance
was even more markedly apparent at its return in
1838. Its bulk, or the actual space occupied by it,
was reduced, as it drew near the hearth of our system
(so far at least as could be inferred from optical
evidence) in the enormous proportion of 800,000 to 1.
A corresponding expansion on each occasion accom-
panied its retirement from the sphere of observation.
Similar changes of volume, though rarely to the same
astounding extent, have been perceived in other comets.
They still remain unexplained ; but it can scarcely be
doubted that they are due to the action of the same
energetic internal forces which reveal themselves in
so many splendid and surprising cometary phenomena." (Hist. Ast., p. 117.) But, as she says on page
115 : "Kepler's remark that comets are consumed by
their own emissions has undoubtedly a measure of
truth in it. The substance ejected into the tail must,
in overwhelmingly large proportion, be forever lost to
the central mass from which it issues. True, it is of
a nature inconceivably tenuous ; but unrepaired waste,
however small in amount, cannot be persisted in with
impunity. The incitement to such self-spoliation
proceeds from the sun; it accordingly progresses
more rapidly the more numerous are the returns to

the solar vicinity. Comets of short period may thus reasonably be expected to wear out quickly."

But while a comet thus loses a part of its mass at each approach to perihelion, it may gain as much or more on its outward and return journey. The "face of the earth" is as a mere point in space compared with the vast area of a comet possessing a small part of the mass of the earth. The amount of meteoric matter gathered by the earth is not sufficient to make an appreciable percentage of increase in the mass of the earth in millions of years; but in actual quantity it is considerable, being estimated by Professor Young at an average somewhere between one ton and one hundred tons per diem. (Young's Genl. Astronomy, art. 777.)

If a comet collects meteoric matter at the same rate over its entire area, enough matter might be added in one orbital journey to make an appreciable difference in both the mass and density of the comet, in the series of years required for its orbital revolution. The increase of mass might be sufficient to retard the return of the comet; and the increase of density might so obstruct the outward flow of the particles surcharged with heliofugal power, on its return to perihelion, as to render a separation or explosion inevitable. Professor Newcomb thus describes the actual disruption of Biela's comet:

"*The Lost Biela's Comet.*—Nothing could more strikingly illustrate the difference between comets and other heavenly bodies than the fact of the total dissolution of one of the former. In 1826, a comet was discovered by an Austrian named Biela, which was found to be

periodic, and to have been observed in 1772, and again
in 1805. The time of the revolution was found to be
six years and eight months. In the next two returns,
the earth was not in the right part of its orbit to admit
of observing the comet; the latter was not therefore
seen again till 1845. In November and December of
that year it was observed as usual, without anything
remarkable being noticed. But in January following,
the astronomers of the Naval Observatory found it
to have suffered an accident never before known to
happen to a heavenly body, and of which no explanation has ever been given. The comet had separated
into two distinct parts, of quite unequal brightness,
so that there were two apparently complete comets,
instead of one. During the month following, the
lesser of the two continually increased, until it became equal to its companion. Then it grew smaller,
and in March vanished entirely, though its companion
was still plainly seen for a month longer. The distance apart of the portions, according to the compu-
tation of Professor Hubbard, was about 200,000
miles." (Pop. Astr., p. 386.) And Miss Clerke adds :
"Biela does not offer the only example of cometary
disruption. Setting aside the unauthentic reports of
early chroniclers, we meet the 'double comet' discovered by Liais at Olinda [Brazil], February 27, 1860, of
which the division appeared recent." (Hist. of Astr.,
p. 387.)

The effect of concentrating the heliofugal power on
one point doubtless causes the end of the comet
toward the sun to become concave, the depth of the
cavity increasing as the comet approaches the sun.

The heat developed in this cavity, into which the direct rays of the sun are steadily poured, must reach a temperature, in comets passing near the sun, of incalculable intensity and a corresponding intensity of centrifugal motion in the heated particles. Under these conditions the disruption or explosion of a comet, as it approaches perihelion, is probably not an unusual occurrence whenever the comet gathers in its orbital journey sufficient meteoric matter to increase its density. If Encke's comet is being retarded by accretions of meteoric matter, we may yet see it split up into two or more fragments, or scattered into meteors and meteorites by an explosion.

The operation of the heliofugal power, as here explained, accounts very satisfactorily for the apparent contraction of the comet's head as it approaches the sun. The matter constituting the head is carried outwards and converted into tail. Professor Young says on this subject:

" *Contraction of a Comet's Head as it approaches the Sun.*—It is a very singular fact that the head of a comet continually and regularly changes its diameter as it approaches to and recedes from the sun; and what is more singular yet, instead of expanding, as one would naturally expect it to do under the action of the solar heat, it contracts when it approaches the sun. No satisfactory explanation is known. . . . The change is especially conspicuous in Encke's comet. When this body first comes into sight, at a distance of about 130,000,000 miles from the sun, it has a diameter of nearly 300,000 miles. When it is near the perihelion, at a distance from the sun of only 33,000,000 miles, its diameter shrinks to 12,000 or 14,000 miles, the volume being then less than $\frac{1}{10000}$ of what it was when first seen ; as it recedes it expands, and resumes its original dimensions. Other comets show a similar but usually less striking change." (Genl. Ast. [Young], art. 715.)

But Professor Steele says : " Comets appear to be subject to constant variations. They are now thought generally to decrease in brilliancy at each successive revolution around the sun. The same comet may present itself sometimes with a tail and sometimes without. When the comet first appears, there is commonly no tail visible and the light is faint. As it approaches the sun, however, its brightness increases, the tail shoots out from the coma, and grows daily in length and splendor. Supernumerary tails, shorter and less distinct than the principal one, dart out, but they generally soon disappear, as if from lack of material. The tail of the comet of 1843, just after the perihelion, increased in length 5,000,000 miles per day. As the tail thus extended, the nucleus was correspondingly contracted, so that this comet actually 'exhausted its head in the manufacture of its own tail.' " (S.'s New Ast., p. 193.)

15. OUTWARD PUSH OF HELIOFUGAL POWER.

It is manifest that the heliofugal power, which causes the violent outward projection of the matter constituting the tail of the comet, must exert an outward push on the entire body of the comet, the push on any part of the cometary matter being proportionate to the amount of heat absorbed. It is this fact which gives the determination of the existence of the heliofugal power an importance far beyond the mere explanation of the cause of the cometary transformation; for it proves that the expansive force of Thermal Repulsion is operative between masses as well as between molecules of a single mass, and that the radiant energy of the sun imparts to matter in space centrifugal or heliofugal, force, proportionate to the quantity of heat absorbed by the matter: in other words, that the correlation of antagonism between Thermal Repulsion and gravitational attraction, which we have seen exists in respect to the molecules of terrestrial matter, exists between the sun and bodies in space.

16. Heliofugal Power causes Planets to revolve.

The obvious effect of this outward push on a solid planet held in its orbit by the counteracting force of gravitation would be to cause the planet to rotate on its axis. The pull of gravitation is from centre to centre of the respective attracting bodies; while the push of Thermal Repulsion is on the surface of the planet exposed to the sun's rays. That side, becoming heated, has imparted to it by the absorption of heat an impetus to get farther from the sun, while the other side, cooled by radiation, acquires an impetus to get nearer the sun. If the planet was stationary, it is possible that the pull and the push might be in equipoise, and no motion result; but the

slightest motion of the planet in its orbit would destroy this equipoise, and give to the heated side the
chance to get away from the sun, and the cooled side
to get nearer to it. This would be effected by an
axial rotation. In the annexed figure let S represent
the sun and P a planet moving in its orbit, without
any axial rotation, and at the point A. At this point
the lighter hemisphere would receive the sun's rays
and the outward impetus, while the darker hemisphere
would be cooled and receive the inward impetus. But
in moving in its orbit without axial rotation from A
to A', a part of the warmed surface would be turned

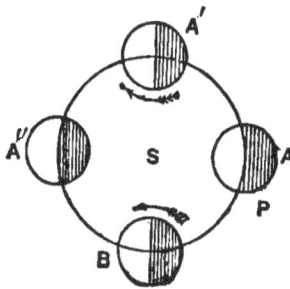

partially from the sun, and a part of the cooled surface
would be turned towards the sun; on making a half-
revolution of its orbit to A'', the hemisphere exposed to
the sun would be the one excluded at A. It is manifest that the outward push of the lighter hemisphere
and the inward pull of the darker hemisphere at A
would cause the planet to revolve the moment it started
to move in its orbit, the rotation being on the inside
and not the outside of its orbit; that is to say, if the orbital motion was from A to A', the axial rotation would
be in the direction of the arrow at A'; if the orbital

motion was from *A* to *B*, the axial revolution would be in the direction of the arrow at *B*. This is precisely the mode in which the earth rotates on its axis; and so far as is known, the other planets all revolve in the same way. The axial revolution causes an equal exposure of the surface longitudinally to the sun's rays, and consequently an equable distribution of the heat. If the earth were held still for a short time in its orbit, or if the same side of it was held continually towards the sun, as the same side of the moon is towards the earth, the inevitable result would be that the atmosphere on the side towards the sun would be heated up and driven off by heliofugal power, like the gaseous matter of a comet's tail, the waters would be vaporized and driven off in the same way, and the solid earth would literally burn up. The intense heat would destroy all living things in a very short time, even before the heliofugal power had carried away the atmosphere.

The intensity of the force of thermal energy is inversely as the square of the distance: as a body approaches the sun, the pull inwards of gravitation and the push outwards of Thermal Repulsion both increase in the same ratio; and as a body recedes from the sun, they each diminish in the same ratio. The time of the earth's axial revolution is practically uniform; the astronomers agree that it certainly has not changed so much as one-tenth of a second in 2000 years. It is evident that the force causing axial rotation is counteracted by some other force; otherwise, the motion, instead of being constantly uniform, would be constantly accelerated. And it is certainly probable that in the

perfect equipoise of Thermal Repulsion and gravitational attraction at all points in the earth's orbit will
be found the explanation of the uniformity of axial
rotation.

17. DIFFERENCE IN SPEED OF AXIAL ROTATION.

Astronomy at present has no satisfactory theory as
to the cause of the axial revolution of the earth and
the other planets, and of course offers no satisfactory
explanation of the wide difference in the time required
for axial revolution in different planets. The earth
revolves in 24 hours; Venus, which is about one-sixth
less than the earth, revolves in 23 hours and 21 minutes; Mars, which is only about one-seventh of the
earth in size, revolves in 24 hours, 37 minutes, 22.7
seconds; while Jupiter, whose mass is nearly 316
times that of the earth, rotates in about 9 hours, 55
minutes; and Saturn, whose mass is about 95 times
that of the earth, revolves in about 10 hours, 14 minutes.* This great variety, which cannot possibly have
any reference to the size of the planet, or its distance
from the sun, may well result from the difference in
the capacity of the several planets to absorb the sun's
heat. If the motion results from the outward push of
the sun's radiant energy, and this push is proportionate to the amount of heat absorbed and the relative attraction of gravitation, it may well be that the
fluffy mass of Jupiter enables him to outstrip all the

---

* These periods are taken from Young's " General Astronomy,"
and are generally accepted ; but some of them have not yet been
fully verified.

other planets in this respect—even the still more fluffy Saturn, because Saturn is doubtless to a great extent sheltered by his rings from the effect of the sun's rays.

18. NOTHING CAN FALL INTO THE SUN.

Another very important result of the determination of the existence of the heliofugal power is that it demonstrates that a body in space cannot fall into the sun. The sun's repellent energy increasing inversely as the square of the distance, and being also proportionate to the heat imparted, it is manifest that any known substance would be vaporized and driven off before reaching the surface of the sun. The danger when a body approaches the sun is not that it will be drawn in and consumed, but that it will be vaporized and blown into driblets through space.

19. EFFECT OF HELIOFUGAL POWER ON ORBITAL MOTION.

This leads us to consider the operation of this force in the orbital motion of comets and planets. It will be observed that, as the repellent force increases and diminishes in precisely the same ratio as the force of gravitation, all calculations of planetary orbital motion based on the attraction of gravitation, an initial velocity, and the mechanical law of action and reaction will remain absolutely unaffected. If we suppose a body moving in space to be acted upon by the inward pull of the sun's attraction of gravitation, and also by the outward push of the radiant energy of the sun, each force increasing and diminishing in the same ratio as the body approaches or recedes from the sun, the resultant orbital motion would be precisely that which would result from the elements now used in calculating orbital motion.

It is obvious from the law of inertia and Newton's first law of motion that circular motion—in fact, any motion except rectilinear—must be the result of two or more forces. Newton demonstrated, and it has been abundantly verified since, that gravitation is one of the forces operative in producing planetary orbital motion; but we have had heretofore nothing better than theory as to the other force or forces which we know must be operative in causing orbital motion. If our induction is correct, it is now manifest that the outward push of the sun's radiant energy on the planet is an element in causing planetary orbital motion; and when the existence of the law of Thermal Repulsion is demonstrated, astronomers will doubtless find in this outward push an explanation of many of the phenomena of orbital motion, the causes of which are not now understood.

All bodies in the solar system, including the sun, exert the force of gravitation in proportion to mass; but in the exertion of Thermal Repulsion there is no such proportion between the other bodies and the sun. It follows that, while the chief pull-in and the chief push-out are from the sun, the planetary bodies are being constantly subjected to strong pulls and but slight pushes from each other. It is not impossible that this inequality in push and pull may be found to account for various phenomena of orbital motion. If the push and the pull were the same, the orbital motion would doubtless be as constant and unvarying as the axial motion of the planets.

The outward push from Thermal Repulsion, like the inward pull of gravitation, is inversely as the square of the distance : consequently, the southern hemisphere of the earth, exposed to the sun at perihelion, receives an outward push stronger than that received by the northern hemisphere exposed at aphelion. This difference in outward push to which the respective poles of the earth are subjected may well account for the change in the direction of the earth's polar axis, which causes the phenomenon of the precession of the equinoxes. It is certainly not less plausible than the accepted hypothesis as to the cause of the phenomenon. It seems not impossible that in the inward pull of gravitation, the outward push of Thermal Repulsion, and the law of inertia will be found the elements which cause a planet, revolving on its axis, to move in an elliptical orbit, with an obliquity of the plane of its ecliptic proportionate to the ellipticity of its orbit and its speed of axial revolution.

The hypothesis that the acceleration of the moon's motion results from an increase in the moon's capacity to absorb the sun's heat is also worthy of consideration. The capacity of a planetary body like the moon to absorb heat is much greater with an atmosphere than without it : the matter lost by comets in passing perihelion doubtless consists largely of hydrogen and hydro-carbons ; and it seems to be conceded that it is this matter which constitutes meteors, which are so numerous in space, the solid matter, more rarely thrown off, constituting the meteorites. The moon must gather meteors as rapidly as the earth, in proportion to size; and the hydrogen and other gaseous matter striking the moon is doubtless at once converted into an atmosphere enveloping the solid body. The moon has not yet gathered enough probably to fill the depressions in its surface,—where this atmosphere would first accumulate,—but the amount gathered annually may be an appreciable percentage of the small quantity previously existing, and may be sufficient to increase the moon's capacity to absorb heat.

If our moon should ever be so fortunate as to come in contact with a gaseous comet, as the moons of Jupiter once did, the effect of gravitation would be to cause the gaseous matter of the comet to wrap itself around the moon, just as our atmosphere enwraps the solid earth. After such an event the amount of heat absorbed by the moon would be greatly increased, especially if any of the gaseous matter was aqueous vapor; and we would doubtless witness a great acceleration in the moon's motion. The moon appears now to be in the condition that the earth was before the formation of stratified rocks. There is no evidence that the earth during that period had an atmosphere. The indications are that it had none; and it is certainly possible that the earth in the ages after the igneous period gathered its atmosphere from meteoric and cometary matter.

But we must leave it to the astronomers to determine the operative effect of Thermal Repulsion in producing orbital motion. Two or more forces are required to produce it, and the inquiring mind cannot rest satisfied with Professor Young's statement that "The mere uniform rectilinear motion of a material mass in space implies no physical cause, and demands no explanation." (General Ast., art. 400.)

This is an assumption that matter was created with the accident of rectilinear motion, and it cannot be accepted as scientifically true while we remain in absolute ignorance of the nature of creative action.

20. THE AMOUNT OF THERMAL ENERGY EXERTED BY THE SUN.

The radiant energy or heliofugal power exercised on the earth's surface by the sun seems to be ample to account for its influence in both axial and orbital motion. Miss Clerke states, as the result of all the calculations on the subject, "that the heat reaching the outskirts of our atmosphere is capable of doing the work of one horse-power for each square yard of the earth's surface." (Clerke's Hist. of Ast., p. 271.)

This amounts to over three million horse-power for each square mile of surface, and, for the whole surface, a force practically inconceivable. It will doubtless be found, by actual calculation, that the outward push of this immense force is the exact equivalent of the inward pull of the sun's attraction.

21. HYPOTHESIS AS TO ORBITS OF COMETS.

All planetary bodies (both comets and planets) are constantly absorbing heat from the sun, and losing heat by radiation. The absorption is least at aphelion and greatest at perihelion. The radiation may be constant, but it is probably greatest at aphelion and least at perihelion. It is therefore obvious that a body pushed outwards by Thermal Repulsion, or heliofugal power, will reach a point in its orbit where the loss of heat by radiation will render it unable to resist the attraction of gravitation. At this point, aphelion, it will return toward the sun; and the nearness of its approach to the sun will depend upon the impetus of its fall. This near approach will cause a more rapid and intense absorption of the sun's heat, and, consequently, a more violent outward push. This seems to be the case with comets, the ellipticity of the orbit having an apparent relation to the nearness of the comet's approach to the sun at perihelion. But unless we knew more about the relative capacity of different comets to absorb heat, which of course depends on the constituent elements of each comet, the foregoing proposition must remain a mere hypothesis.

22. HELIOFUGAL POWER IN THE EARTH'S ATMOSPHERE— TRADE-WINDS.

The effect of this heliofugal power on the earth's atmosphere explains many meteorological phenomena, and furnishes a basis for more exactness than has heretofore been possible in the science of meteorology. If centrifugal or heliofugal motion is imparted to matter by the absorption of the sun's heat, it is obvious that aqueous vapor must be put in motion by the heat absorbed in evaporation. The direction of the force is of course away from the sun; and the resultant motion would be in that direction also if not resisted, and, if resisted, it would be in the direction of least resistance. The motion is resisted by the atmosphere, into which the aqueous vapor is diffused, and by the solid earth; but in the equatorial regions and over the ocean, where the evaporation is greatest, the motion of the aqueous vapor appears to be strong enough to put the atmosphere in motion, causing the trade-winds. These winds blow steadily from northeast to southwest in the northern hemisphere, and from southeast to northwest in the southern hemisphere. They are confined to an area on each side of the thermal equator, where the solar evaporation is greatest; the annual evaporation in the trade-wind region being from 11 to 12 feet of the ocean surface. The general direction is away from the sun, and the steadiness of the motion shows a like steadiness in the force causing the motion. The accepted theory that the trade-winds result from convection, is not sustained by the facts. On the land within the tropics, heating is greater than on the sea, and there is necessarily more convection: but there is no trade-wind. If convection caused these winds, the air at the polar edges of the trade-winds would be cold; but this is not the fact. The air supposed by the theory to be coming in to take the place of the heated air raised up by convection is itself warm.

23. HELIOFUGAL POWER CAUSING CYCLONES.

We have also in the heliofugal energy imparted to aqueous vapor in the process of evaporation a probable explanation of the fall of the barometer on the approach of a storm, and of the fact that the most violent atmospheric motion is always when the air is most thoroughly saturated with aqueous vapor. We have seen that the heliofugal power imparts motion to the superheated matter of the comet, causing its outward projection—this outward projection, in respect to the other parts of the comet, being proportionate to the difference in the heat absorbed by the different parts of the comet. If this same phenomenon occurs in respect to the formation of aqueous vapor on the earth's surface, the motion thus imparted must continue until successfully resisted, or until it is destroyed by condensing the vapor.

Professor Ferrell says : " As the power of the cyclone is mostly in the aqueous vapor condensed, and without this we rarely have the conditions of more than an initial cyclonic action, the velocity and direction of the progressive motion of a cyclone depends, to some extent at least, upon the distribution of this vapor in the region in which the cyclone exists, and the cyclone is likely to be drawn somewhat in the direction in which there is the most vapor. For this reason it is, perhaps, that the chain of lakes between Canada and the United States seems to be a great highway for cyclones." (Report Chief Sig. Ser. Off. U. S., 1885, part 2, p. 260.) It is not impossible that in this force and its resultant motion will be found an explanation of the phenomenon that the vortex motion of whirlpools and cyclones in the northern hemisphere is in one direction, and in the southern hemisphere in the opposite direction ; and that vines twine around poles and bodies of trees in the same direction as the vortex motion in each hemisphere. The accepted theory that this difference in the direction of vortex motion results from the earth's rotation is not satisfactory, and, like the trade-wind theory, seems to have very little reference to observed facts.

But this force, causing motion in the atmosphere, is necessarily inextricably intertangled with the force of convection, and it will require extended examination to determine to what extent any wind is the result of the one or the other of these forces.

24. LABORATORY EXPERIMENTS.

It will be remembered that, in determining the existence of the law from the phenomena of terrestrial matter, the only question not fully settled was whether the expansive energy of heat operated between masses of terrestrial matter as it does between the molecules of a single mass. It is not impossible that a laboratory demonstration may be found in a practical analysis of the phenomenon of convection. This term is used to describe the motion which results when heat is developed underneath or within a liquid or gas : the heated particles rising upwards, while the particles not heated sink downwards to supply the place of those which rise up. This circulatory motion always occurs when the heat is applied at the bottom of a vessel containing a liquid or gas ; and in such cases the liquid or gas acquires and maintains a uniform temperature throughout the mass. But the motion does not result if the heat is applied at the top, and, consequently, in heating water, the heat is applied at the bottom of the vessel containing the water.

These well-known facts have led to the assumption that convection resulted from the difference in the specific gravity of the heated particles and the particles not heated; and this is undoubtedly one cause of it. But is this the only cause? The motion in the atmosphere in some cases appears to be greater than we ought to expect from mere difference in specific gravity, suggesting the possibility that this force is supplemented by another. The difference in specific gravity itself results from the fact that the heat pushes apart the molecules constituting the portion heated, thus forcing the same mass to occupy a larger volume of space. This pushing out of the molecules necessarily overcomes, to the extent of the expansion, the gravitational attraction of the molecules for each other; and there appears to be no reason why the same force would not affect the gravitational attraction between the earth and each one of these molecules. The attraction of the earth is so immense, when compared to the attraction of a molecule, that it may be impossible to observe the difference in gravitational attraction between the earth and a molecule of liquid or gas caused by the application of heat; but it is impossible to resist the conviction that there is such a difference, because we see the difference in gravitational attraction between the molecules themselves in the phenomenon of expansion.

Our knowledge of the uniformity of nature leaves no room for doubt that the heliofugal power, which causes the outward projection of the tenuous matter composing the comet's tail, is exerted whenever the sun shines on similar matter. The matter constituting the earth's surface and its atmosphere is not essentially different from other planetary and cometary matter; and we therefore know that this heliofugal force is operative on terrestrial matter, and can be found if we look close enough for it. The operation of the gravitational pull between masses of matter is easily understood; but the same pull between the molecules of a separate mass is obscure, and some scientific people yet find it difficult to believe that shot-makers use this force in making drop-shot, and that it is operative in the formation of a dew-drop. On the other hand, the operation of the push of Thermal Repulsion between molecules is obvious in the ordinary phenomena of expansion; but the existence of this push between masses of matter is even more obscure than the intermolecular pull of gravitation. We have nothing like shot-making or the dewdrop to illustrate its operation; but the phenomena of comets demonstrate the operation of this push of Thermal Repulsion between cosmic masses, and it ought to be found operative between masses of terrestrial matter. A torsion balance such as Bailey used to determine the gravitational pull between masses, operated in vacuum, could probably be made to show the extent that the gravitational pull between masses can be counteracted by the push of Thermal Repulsion.

The reason why the existence of the force has heretofore escaped attention is that its operation in all experiments in the atmosphere is necessarily identical with that of convection and increased buoyancy from expansion. The application of heat always causes

upward motion in the atmosphere, from convection:
the heated body expands, occupies more space in the
atmosphere than a similar body not heated, and therefore weighs less. The direct diminution of gravitational pull is completely occluded under these two
forces, and is therefore not observed in the most deli-
cate experiments conducted in the atmosphere.

25. MR. CROOKES' EXPERIMENTS.

The experiments of Mr. Crookes came very near
demonstrating the existence of this force in terrestrial
matter. From the vague and unsatisfactory accounts of
his early experiments which I have been able to obtain,
it seems that he observed or suspected a diminution
of the earth's gravitational attraction from the application of heat to a mass in vacuum. This diminution
would necessarily be almost infinitesimal, in respect to
the small masses which could be weighed in vacuum,
because of the immensity of the earth's pull. But between separate masses of matter the gravitational pull
is very small, compared to the earth's pull on each of
them; and consequently the diminution of the pull
between them by the push of Thermal Repulsion ought
to be more obvious. It seems to me, therefore, probable that the motion of the fly in the radiometer is, in
part at least, the direct result of the force of radiant
energy, as at first supposed by Mr. Crookes himself.
His subsequent conclusion, that the motion is the result of the impact of the molecules of the residual gas,
may be true also, for there is nothing inconsistent in
the two forces. The assumption that the fly will not

move at all in absolute vacuum will probably never be verified by actual experiment.

26. General Remarks.

It has not been deemed expedient in this inquiry to discuss the various theories as to the nature of heat—its development, supply, and the like. "Not the slightest variation has been detected in the amount of heat received from the sun." (Newcomb's Pop. Astronomy, p. 511.) And Professor Young says: "It is an interesting and thus far unsolved problem whether the total amount of the sun's radiation varies perceptibly at different times. It is only certain that the variations, if real, are too small to be detected by our present means of observation. Possibly at some time in the future, observation on a mountain summit above the main body of our atmosphere will decide the question. Since the Christian era, however, it is certain that the amount of heat annually received from the sun has remained practically unchanged. This is inferred from the distribution of plants and animals, which is still substantially the same as in the days of Pliny." (Genl. Ast., art. 352.)

The force of gravitation, and its continued activity, require explanation as much as the phenomena of heat and its supply. Our knowledge of the origin and source of supply of each of these forces is at present a perfect blank. We can develop heat, but it has been demonstrated that this is done by converting some other form of energy, potential or actual, into heat. We can no more create heat or electricity than we

can create matter or gravitation ; and there is no phe-
nomenal evidence that Nature is undergoing an increase or diminution of any one of these elements.
We know certainly that we know absolutely nothing
about the origin of any of them ; and in this density of
ignorance speculation will not reach the truth.

27. CONCLUSION.

The well-known phenomena of nature which we
have been considering demonstrate that there is an
essential difference between matter and force in the
Constitution of Nature ; that force is not in one form,
but in many forms, and that two of these forms or
manifestations of force, heat and gravitation, are ever
present and in active operation where matter exists ;
that these forces operate on the molecules constituting a separate mass of matter, the force of gravitation being a pull inwards towards the centre of
mass, and the force of heat being a push outwards
from the centre ; that outward and inward motion
of the molecules is the result of the predominance
of the one or the other of these forces, and that
the motion (contraction or expansion) is uniform, except when intercepted by some other force ; that the
inward pull of gravitation between separate masses of
matter is identically the same as the pull between the
molecules of a single mass ; and that, while it has not
yet been fully demonstrated, we are justified in as-
suming that the outward push of heat is the same
between separate masses of matter as between the
molecules of a single mass,—this being true, it follows

that all matter in Nature is held suspended between these two forces of attraction and repulsion. Within the earth itself Nature has stored up heat more than ample to reduce all forms of matter to the most tenuous gas, and the immense outward push of this vast self-acting boiler counteracts the inward pull of gravity; and thus it is, that Thermal Repulsion and Gravitational Attraction hold in position the very ground beneath our feet. The end of the world, as we know it, would come by an explosion or contraction, if either of these forces was suspended for an instant.